MY SENSES

Touch

CHRISTINA EARLEY

A Crabtree Roots Book

CRABTREE
Publishing Company
www.crabtreebooks.com

School-to-Home Support for Caregivers and Teachers

This book helps children grow by letting them practice reading. Here are a few guiding questions to help the reader with building his or her comprehension skills. Possible answers appear here in red.

Before Reading:

- What do I think this book is about?
    - *I think this book is about my sense of touch.*
    - *I think this book is about things I can touch.*

- What do I want to learn about this topic?
    - *I want to learn about the part of my body I use for touch.*
    - *I want to learn about different things I can touch.*

During Reading:

- I wonder why...
    - *I wonder why my skin feels cold in the pool.*
    - *I wonder why my skin feels hot in the Sun.*

- What have I learned so far?
    - *I have learned that I use my skin to touch.*
    - *I have learned that tree bark is rough.*

After Reading:

- What details did I learn about this topic?
    - *I have learned that touch is one of my five senses.*
    - *I have learned that I should not touch very hot things.*

- Read the book again and look for the vocabulary words.
    - *I see the word **rough** on page 7 and the word **smooth** on page 8. The other vocabulary words are found on page 14.*

Touch is one of my five **senses**.

My **skin** tells me how things feel.

This chick is **soft**.

This bark is **rough**.

This rock is **smooth**.

The Sun feels hot on my skin.

The pool feels cold and wet.

I am **careful** what I touch.

Word List

Sight Words

how	of	things
is	on	this
me	one	
my	the	

Words to Know

careful

rough

senses

skin

smooth

soft

45 Words

Touch is one of my five **senses**.

My **skin** tells me how things feel.

This chick is **soft**.

This bark is **rough**.

This rock is **smooth**.

The Sun feels hot on my skin.

The pool feels cold and wet.

I am **careful** what I touch.

Written by: Christina Earley
Designed by: Rhea Wallace
Series Development: James Earley
Proofreader: Janine Deschenes
Educational Consultant: Marie Lemke M.Ed.

Photographs:
Shutterstock: Eugene Kochera: cover; A3pfamily: p. 3, 14; Ann in the U.K.: p. 4, 14; CHEN MIN CHUN: p. 5, 14; lakshmiprasada S: p. 6, 14; Matveev Aleksandr: p. 9, 14; vvvita: p. 10; ESB Professional: p. 11; Lamyai: p. 13, 14; BlurryMe: p. 13(bottom)s

Library and Archives Canada Cataloguing in Publication

Available at the Library and Archives Canada

Library of Congress Cataloging-in-Publication Data

Available at the Library of Congress

Crabtree Publishing Company
www.crabtreebooks.com   1-800-387-7650

Printed in the U.S.A./062021/CG20210401

Copyright © 2022 **CRABTREE PUBLISHING COMPANY**

All rights reserved. No part of this publication may be reproduced, stored in a retrieval system or be transmitted in any form or by any means, electronic, mechanical, photocopying, recording, or otherwise, without the prior written permission of Crabtree Publishing Company. In Canada: We acknowledge the financial support of the Government of Canada through the Canada Book Fund for our publishing activities.

Published in the United States
Crabtree Publishing
347 Fifth Avenue, Suite 1402-145
New York, NY, 10016

Published in Canada
Crabtree Publishing
616 Welland Ave.
St. Catharines, Ontario L2M 5V6